INTERNATIONAL STRATEGY
FOR CYBERSPACE

Prosperity, Security, and Openness
in a Networked World

MAY 2011

THE WHITE HOUSE
WASHINGTON

Cyberspace, and the technologies that enable it, allow people of every nationality, race, faith, and point of view to communicate, cooperate, and prosper like never before. Today, an American company can do business anywhere in the world with an Internet connection, supporting countless jobs and opportunities for the American people. A mother in rural Africa can sell crafts to a family in Latin America, advancing broader economic development. A laboratory in Europe can conduct field-changing research on hardware made in Asia and software written in North America, and students in Australia and the Middle East can learn together through videoconference. And more than ever, citizens across the globe are being empowered with information technologies to help make their governments more open and responsive.

Today, as nations and peoples harness the networks that are all around us, we have a choice. We can either work together to realize their potential for greater prosperity and security, or we can succumb to narrow interests and undue fears that limit progress. Cybersecurity is not an end unto itself; it is instead an obligation that our governments and societies must take on willingly, to ensure that innovation continues to flourish, drive markets, and improve lives. While offline challenges of crime and aggression have made their way to the digital world, we will confront them consistent with the principles we hold dear: free speech and association, privacy, and the free flow of information.

The digital world is no longer a lawless frontier, nor the province of a small elite. It is a place where the norms of responsible, just, and peaceful conduct among states and peoples have begun to take hold. It is one of the finest examples of a community self-organizing, as civil society, academia, the private sector, and governments work together democratically to ensure its effective management. Most important of all, this space continues to grow, develop, and promote prosperity, security, and openness as it has since its invention. This is what sets the Internet apart in the international environment, and why it is so important to protect.

In this spirit, I offer the United States' International Strategy for Cyberspace. This is not the first time my Administration has addressed the policy challenges surrounding these technologies, but it is the first time that our Nation has laid out an approach that unifies our engagement with international partners on the full range of cyber issues. And so this strategy outlines not only a vision for the future of cyberspace, but an agenda for realizing it. It provides the context for our partners at home and abroad to understand our priorities, and how we can come together to preserve the character of cyberspace and reduce the threats we face.

By itself, the Internet will not usher in a new era of international cooperation. That work is up to us, its beneficiaries. Together, we can work together to build a future for cyberspace that is open, interoperable, secure, and reliable. This is the future we seek, and we invite all nations, and peoples, to join us in that effort.

Table of Contents

I. Building Cyberspace Policy

"This world—cyberspace—is a world that we depend on every single day... [it] has made us more interconnected than at any time in human history."

—President Barack Obama, May 29, 2009

Digital infrastructure is increasingly the backbone of prosperous economies, vigorous research communities, strong militaries, transparent governments, and free societies. As never before, information technology is fostering transnational dialogue and facilitating the global flow of goods and services. These social and trade links have become indispensable to our daily lives. Critical life-sustaining infrastructures that deliver electricity and water, control air traffic, and support our financial system all depend on networked information systems. Governments are now able to streamline the provision of essential services through eGovernment initiatives. Social and political movements rely on the Internet to enable new and more expansive forms of organization and action. The reach of networked technology is pervasive and global. For all nations, the underlying digital infrastructure is or will soon become a national asset.

To realize fully the benefits that networked technology promises the world, these systems must function reliably and securely. People must have confidence that data will travel to its destination without disruption. Assuring the free flow of information, the security and privacy of data, and the integrity of the interconnected networks themselves are all essential to American and global economic prosperity, security, and the promotion of universal rights.

Almost a third of the world's population uses the Internet and countless more are touched by it in their daily lives. There are more than four billion digital wireless devices in the world today. Scarcely a half-century ago, that number was zero. We live in a rare historical moment with an opportunity to build on cyberspace's successes and help secure its future for U.S. citizens and the global community.

For these technologies to continue to empower individuals, enrich societies, and foster the research, development, and innovation essential to building modern economies, it must retain the openness and interoperability that have characterized its explosive growth. Underlying these are technical principles and effective governance structures that demand our support. At the same time, our networks must be secure and reliable; they must retain the trust of individuals, businesses and governments, and should be resilient to arbitrary or malicious disruption.

The world must collectively recognize the challenges posed by malevolent actors' entry into cyberspace, and update and strengthen our national and international policies accordingly. Activities undertaken in cyberspace have consequences for our lives in physical space, and we must work towards building the rule of law, to prevent the risks of logging on from outweighing its benefits. The future of an open, interoperable, secure and reliable cyberspace depends on nations recognizing and safeguarding that which should endure, while confronting those who would destabilize or undermine our increasingly networked world.

Strategic Approach

The foundation of the United States' international cyberspace policy is the belief that networked technologies hold immense potential for our Nation, and for the world. Over the last three decades we, the United States, have watched these technologies revolutionize our economy and transform of our daily lives. We have also witnessed offline challenges, like exploitation and aggression, move into cyberspace. As we adapt to meet those challenges, we will lead by example. The United States will pursue an international cyberspace policy that empowers the innovation that drives our economy and improves lives here and abroad. In all this work, we are grounded in principles essential not just to American foreign policy, but to the future of the Internet itself.

Building on Successes

The United States is committed to preserving and enhancing the benefits of digital networks to our societies and economies.

These benefits have been diverse and profound. For individuals, computer networks have enhanced productivity and prosperity; helped to overcome disadvantage and disability; brought together those isolated by language or a rare disease; connected families and friends across distant and often-fraught borders. For communities, they have sped first response to emergencies, expanded information-sharing to help solve crimes, shed light on corruption, facilitated political action, and brought wide attention to overlooked causes. For businesses, they have opened new markets and spawned billion-dollar industries. For governments, they have enabled increased transparency, efficiency, and convenience, and have connected leaders to those they serve. For the international community, they have provided the foundation for a new global marketplace of ideas, and helped channel remarkable generosity in the face of tragedy. The more freely information flows, the stronger our societies become. Properly used, these technologies can strengthen us all, and we will work to expand their reach and improve their operation at home and abroad.

Recognizing the Challenges

The United States acknowledges that the growth of these networks brings with it new challenges for our national and economic security and that of the global community.

These challenges come in a variety of forms. Natural disasters, accidents, or sabotage can disrupt cables, servers, and wireless networks on U.S. soil and beyond. Technical challenges can be equally disruptive, as one country's method for blocking a website can cascade into a much larger, international network disruption. Extortion, fraud, identity theft, and child exploitation can threaten users' confidence in online commerce, social networks and even their personal safety. The theft of intellectual property threatens national competitiveness and the innovation that drives it. These challenges transcend national borders; low costs of entry to cyberspace and the ability to establish an anonymous virtual presence can also lead to "safe havens" for criminals, with or without a state's knowledge. Cybersecurity threats can even endanger international peace and security more broadly, as traditional forms of conflict are extended into cyberspace.

Grounded in Principle

The United States will confront these challenges—while preserving our core principles.

Our policies flow from a commitment to both preserving the best of cyberspace and safeguarding our principles. Our international cyberspace policy reflects our core commitments to *fundamental freedoms, privacy,* and the *free flow of information.*

Fundamental Freedoms. Our commitment to freedom of expression and association is abiding, but does not come at the expense of public safety or the protection of our citizens. Among these civil liberties, recognized internationally as "fundamental freedoms," the ability to seek, receive and impart information and ideas through any medium and regardless of frontiers has never been more relevant. As a nation, we are not blind to those Internet users with malevolent intentions, but recognize that exceptions to free speech in cyberspace must also be narrowly tailored. For example, child pornography, inciting imminent violence, or organizing an act of terrorism have no place in any society, and thus, they have no place on the Internet. Nonetheless, the United States will continue to combat them in a manner consistent with our core values—treating these issues specifically, and not as referenda on the Internet's value to society.

Privacy. Our strategy marries our obligation to protect our citizens and interests with our commitment to privacy. As citizens increasingly engage via the Internet in their public and private lives, they have expectations for privacy: individuals should be able to understand how their personal data may be used, and be confident that it will be handled fairly. Likewise, they expect to be protected from fraud, theft, and threats to personal safety that lurk online—and expect law enforcement to use all the tools at their disposal, pursuant to law, to track and prosecute those who would use the Internet to exploit others. The United States is committed to ensuring balance on both sides of this equation, by giving law enforcement appropriate investigative authorities it requires, while protecting individual rights through appropriate judicial review and oversight to ensure consistency with the rule of law.

Free Flow of Information. States do not, and should not have to choose between the free flow of information and the security of their networks. The best cybersecurity solutions are dynamic and adaptable, with minimal impact on network performance. These tools secure systems without crippling innovation, suppressing freedom of expression or association, or impeding global interoperability. In contrast, we see other approaches—such as national-level filters and firewalls—as providing only an illusion of security while hampering the effectiveness and growth of the Internet as an open, interoperable, secure, and reliable medium of exchange. The same is true commercially; cyberspace must remain a level playing field that rewards innovation, entrepreneurship, and industriousness, not a venue where states arbitrarily disrupt the free flow of information to create unfair advantage. The United States is committed to international initiatives and standards that enhance cybersecurity while safeguarding free trade and the broader free flow of information, recognizing our global responsibilities, as well as our national needs.

Too often, such principles are characterized as incompatible with effective law enforcement, anonymity, the protection of children and secure infrastructure. In reality, good cybersecurity can enhance privacy, and effective law enforcement targeting widely-recognized illegal behavior can protect fundamental freedoms. The rule of law—a civil order in which fidelity to laws safeguards people and interests; brings stability to global markets; and holds malevolent actors to account internationally—both supports our national security and advances our common values.

II. Cyberspace's Future

Envision a future in which reliable access to the Internet is available from nearly any point on the globe, at a price that businesses and families can afford. Computers can communicate with one another across a seamless landscape of global networks permitting trusted, instantaneous communication with friends and colleagues down the block or around the world. Content is offered in local languages and flows freely beyond national borders, as improvements in digital translation open to millions a wealth of knowledge, new ideas, and rich debates. New technologies improving agriculture or promoting public health are shared with those in greatest need, and difficult problems benefit from global collaboration among experts and innovators. This, in part, is the future of cyberspace that the United States seeks—and the future we will work to realize.

In this future, individuals and businesses can quickly and easily obtain the tools necessary to set up their own presence online; domain names and addresses are available, secure, and properly maintained, without onerous licenses or unreasonable disclosures of personal information. The best engineers work together internationally to develop new standards for information systems that make networks faster and more reliable, catalyzing innovation and expanding accessibility. High-tech industry works with its customers to provide software, hardware, and services that are more secure, more reliable, and more responsive to their needs.

It is a future in which universities and companies are free to research and develop new concepts and products because they know their intellectual property and valuable data are safe, even on shared networks. Individuals know the threats to their personal computers, and can take easy-to-use measures to protect their systems. Private-sector companies also take a responsibility for their network hygiene, knowing that in so doing, they protect their investments. When cybersecurity incidents demand government action, officials can detect those threats early and share data in real-time to mitigate the spread of malware or minimize the impact of a major disruption—all while preserving the broader free flow of information. When a crime is committed internationally, law enforcement agencies are able to collaborate to safeguard and share evidence and bring individuals to justice.

This future promises not just greater prosperity and more reliable networks, but enhanced international security and a more sustainable peace. In it, states act as responsible parties in cyberspace—whether configuring networks in ways that will spare others disruption, or inhibiting criminals from using the Internet to operate from safe havens. States know that networked infrastructure must be protected, and they take measures to secure it from disruption and sabotage. They continue to collaborate bilaterally, multilaterally, and internationally to bring more of the world into the information age and into the consensus of states that seek to preserve the Internet and its core characteristics.

The United States and a growing number of partners have laid the foundation for this future already. But it is not a foregone conclusion, and we cannot build it alone. Though progress may be slow and resource-intensive, the international community must join together in support of this long-term investment. We must do so with the clear understanding that this vision of cyberspace serves national interests as much as shared international aims. The measure of our success will be another half-century of information technology as transformational as the last, as we begin to realize fully the benefits—and minimize the risks—of global interconnection.

The Future We Seek

The cyberspace environment that we seek rewards innovation and empowers individuals; it connects individuals and strengthens communities; it builds better governments and expands accountability; it safeguards fundamental freedoms and enhances personal privacy; it builds understanding, clarifies norms of behavior, and enhances national and international security. To sustain this environment, international collaboration is more than a best practice; it is a first principle.

Our Goal

The United States will work internationally to promote an **open, interoperable, secure, and reliable** information and communications infrastructure that supports international trade and commerce, strengthens international security, and fosters free expression and innovation. To achieve that goal, we will build and sustain an environment in which **norms of responsible behavior** guide states' actions, sustain partnerships, and support the rule of law in cyberspace.

Open and Interoperable: A Cyberspace That Empowers

At the core of digital innovation is the ability to add new functionality to networked machines. The openness of digital systems explains their explosive growth, rapid development, and enduring importance. Networked technology's basic tools are steadily increasing in availability and decreasing in price, as computer and Internet access have spread to every nation. To continue to serve the needs of an ever-growing wired population, manufacturers of hardware and operating systems must continue to empower the widest possible range of developers across the globe. As companies continue to drive innovation in the development of proprietary software, we also applaud the vibrancy of the open-source software movement, giving developers and consumers the choice of community-driven solutions to meet their needs.

The United States supports an Internet with end-to-end interoperability, which allows people worldwide to connect to knowledge, ideas, and one another through technology that meets their needs. The free flow of information depends on interoperability—a principle affirmed by 174 nations in the Tunis Commitment of the World Summit on the Information Society. The alternative to global openness and interoperability is a fragmented Internet, where large swaths of the world's population would be denied access to sophisticated applications and rich content because of a few nations' political interests. The collaborative development of consensus-based international standards for information and communication technology is a key part of preserving openness and interoperability, growing our digital economies, and moving our societies forward.

Secure and Reliable: A Cyberspace That Endures

For cyberspace as we know it to endure, our networked systems must retain our trust. Users need to have confidence that their data will be secure in transit and storage, as well as reliable in delivery. An effective strategy will require action on many fronts, with shared responsibility at every level of society, from the end-user up through collaboration among nation-states.

Vulnerability reduction will require robust technical standards and solutions, effective incident management, trustworthy hardware and software, and secure supply chains. Risk reduction on a global scale will require effective law enforcement; internationally agreed norms of state behavior; measures that build confidence and enhance transparency; active, informed diplomacy; and appropriate deterrence. Finally, incident response will require increased collaboration and technical information sharing with the private sector and international community. This work cannot be fully addressed by any single nation or sector alone; it is a responsibility and duty that every nation, and its people, all share.

Network stability is a cornerstone of our global prosperity, and securing those networks is more than strictly a technical matter. Economically, we must advance sustainable growth and invest in infrastructure at home and abroad, while incentivizing network reliability and clarifying the obligations of firms and states. Politically, we must help to maintain an environment of respect for technical infrastructure, so disputes do not become excuses to disrupt and degrade networks. Socially, we must make end-users aware of their responsibilities to maintain and operate their devices in a safe and secure manner.

Stability Through Norms

The United States will work with like-minded states to establish an environment of expectations, or norms of behavior, that ground foreign and defense policies and guide international partnerships. The last two decades have seen the swift and unprecedented growth of the Internet as a social medium; the growing reliance of societies on networked information systems to control critical infrastructures and communications systems essential to modern life; and increasing evidence that governments are seeking to exercise traditional national power through cyberspace. These events have not been matched by clearly agreed-upon norms for acceptable state behavior in cyberspace. To bridge that gap, we will work to build a consensus on what constitutes acceptable behavior, and a partnership among those who view the functioning of these systems as essential to the national and collective interest.

The Role of Norms. In other spheres of international relations, shared understandings about acceptable behavior have enhanced stability and provided a basis for international action when corrective measures are required. Adherence to such norms brings predictability to state conduct, helping prevent the misunderstandings that could lead to conflict.

The development of norms for state conduct in cyberspace does not require a reinvention of customary international law, nor does it render existing international norms obsolete. Long-standing international norms guiding state behavior—in times of peace and conflict—also apply in cyberspace. Nonetheless, unique attributes of networked technology require additional work to clarify how these norms apply and what additional understandings might be necessary to supplement them. We will continue to work internationally to forge consensus regarding how norms of behavior apply to cyberspace, with the understanding that an important first step in such efforts is applying the broad expectations of peaceful and just interstate conduct to cyberspace.

The Basis for Norms. Rules that promote order and peace, advance basic human dignity, and promote freedom in economic competition are essential to any international environment. These principles provide a basic roadmap for how states can meet their traditional international obligations in cyberspace and, in many cases, reflect duties of states that apply regardless of context. The existing principles that should support cyberspace norms include:

- **Upholding Fundamental Freedoms:** States must respect fundamental freedoms of expression and association, online as well as off.

- **Respect for Property:** States should in their undertakings and through domestic laws respect intellectual property rights, including patents, trade secrets, trademarks, and copyrights.

- **Valuing Privacy:** Individuals should be protected from arbitrary or unlawful state interference with their privacy when they use the Internet.

- **Protection from Crime:** States must identify and prosecute cybercriminals, to ensure laws and practices deny criminals safe havens, and cooperate with international criminal investigations in a timely manner.

- **Right of Self-Defense:** Consistent with the United Nations Charter, states have an inherent right to self-defense that may be triggered by certain aggressive acts in cyberspace.

Deriving from these traditional principles of interstate conduct are responsibilities more specific to cyberspace, focused in particular on preserving global network functionality and improving cybersecurity. Many of these responsibilities are rooted in the technical realities of the Internet. Because the Internet's core functionality relies on systems of trust (such as the Border Gateway Protocol), states need to recognize the international implications of their technical decisions, and act with respect for one another's networks and the broader Internet. Likewise, in designing the next generation of these systems, we must advance the common interest by supporting the soundest technical standards and governance structures, rather than those that will simply enhance national prestige or political control. Emerging norms, also essential to this space, include:

- **Global Interoperability:** States should act within their authorities to help ensure the end-to-end interoperability of an Internet accessible to all.

- **Network Stability:** States should respect the free flow of information in national network configurations, ensuring they do not arbitrarily interfere with internationally interconnected infrastructure.

- **Reliable Access:** States should not arbitrarily deprive or disrupt individuals' access to the Internet or other networked technologies.

- **Multi-stakeholder Governance:** Internet governance efforts must not be limited to governments, but should include all appropriate stakeholders.

- **Cybersecurity Due Diligence:** States should recognize and act on their responsibility to protect information infrastructures and secure national systems from damage or misuse.

While cyberspace is a dynamic environment, international behavior in it must be grounded in the principles of responsible domestic governance, peaceful interstate conduct, and reliable network management. As these ideas develop, the United States will foster and participate fully in discussions, advancing a principled approach to Internet policy-making and developing shared understandings in fora appropriate to each issue.

Our Role in Cyberspace's Future

To realize this future and help promulgate positive norms, the United States will combine diplomacy, defense, and development to enhance prosperity, security, and openness so all can benefit from networked technology. These three approaches are central to our efforts internationally. In the latter half of the 20th century, the United States helped forge a new post-war architecture of international economic and security cooperation. In the 21st century, we will work to realize this vision of a peaceful and reliable cyberspace in that same spirit of cooperation and collective responsibility.

Diplomacy: Strengthening Partnerships

Extending the principles of peace and security to cyberspace—while preserving its benefits and character—will require strengthened partnerships and expanded initiatives. We will engage the international community in frank and urgent dialogue, to build consensus around principles of responsible behavior in cyberspace and the actions necessary, both domestically and as an international community, to build a system of cyberspace stability.

Diplomatic Objective

The United States will work to create incentives for, and build consensus around, an international environment in which states—recognizing the intrinsic value of an open, interoperable, secure, and reliable cyberspace—work together and act as responsible stakeholders.

Strengthening Partnerships

Through our international relationships and affiliations, we will seek to ensure that as many stakeholders as possible are included in this vision of cyberspace precisely because of its economic, social, political, and security benefits. These efforts will be supported by meaningful collaboration with the private sector at home and abroad.

Distributed systems require distributed action, and no single institution, document, arrangement, or instrument could suffice in addressing the needs of our networked world. From end-users, private-sector hardware and software vendors, and Internet service providers, to regional, multilateral, and multi-stakeholder organizations—all are important in helping cyberspace meet its full potential.

In the international arena in particular, states have an enduring role to play in preserving peace and stability, empowering innovation, safeguarding economic and national security interests, and protecting and promoting the individual rights of citizens. In our international relations, the United States will work to establish an environment of international expectations that anchor foreign and defense policies and strengthen our international relationships.

Bilateral and Multilateral Partnerships. We will work bilaterally with nations to build collaboration on cyberspace issues important to our governments and our peoples. Building broad international understanding about cyberspace norms of behavior must begin with clear agreement among like-minded countries. We will seek a broad community of partners in these efforts, and will include cyberspace issues in a wide range of bilateral dialogues, at all levels of government and across a wide range of our activities. We will advance common action on cyberspace's emerging challenges, while building on those enforcement tools and approaches already enjoying success. Furthermore, we will actively engage the developing world, and ensure that emerging voices on these issues are heard.

International and Multi-stakeholder Organizations. Regional organizations have been particularly effective at tackling cybersecurity problems specific to their members. They will play an increasingly important role in developing and applying norms of behavior. We will continue to use our membership in these organizations, as well as in broader international organizations, to develop productive agendas that are appropriate to each organization's expertise and that realize concrete benefits for members. In Internet governance policy, important steps have been made to ensure responsiveness and international representation in key organizations. The United States salutes those efforts, and will continue to recognize the unique contribution of such fora that represent the entire Internet community by integrating the private sector, civil society, academia, as well as governments in a multi-stakeholder environment.

Private Sector Collaboration. Although the private sector already plays an important role in international and multi-stakeholder organizations, we will continue to leverage existing partnership mechanisms to engage with industry partners. In particular, we will work closely with infrastructure owners and operators—who are responsible for the majority of network functionality—to expand initiatives to secure the network ecosystem, preserve the benefits and character of cyberspace, avoid unnecessary impediments to technological evolution, and extend principles of peace and security. We also seek the private sector's participation in Internet governance as essential to upholding its multi-stakeholder character, and will continue to advocate for inclusiveness in fora that take up such issues.

Defense: Dissuading and Deterring

The United States will defend its networks, whether the threat comes from terrorists, cybercriminals, or states and their proxies. Just as importantly, we will seek to encourage good actors and dissuade and deter those who threaten peace and stability through actions in cyberspace. We will do so with overlapping policies that combine national and international network resilience with vigilance and a range of credible response options. In all our defense endeavors, we will protect civil liberties and privacy in accordance with our laws and principles.

Defense Objective

The United States will, along with other nations, encourage responsible behavior and oppose those who would seek to disrupt networks and systems, dissuading and deterring malicious actors, and reserving the right to defend these vital national assets as necessary and appropriate.

Dissuasion

Protecting networks of such great value requires robust defensive capabilities. The United States will continue to strengthen our network defenses and our ability to withstand and recover from disruptions and other attacks. For those more sophisticated attacks that do create damage, we will act on well-developed response plans to isolate and mitigate disruption to our machines, limiting effects on our networks, and potential cascade effects beyond them.

Strength at Home. Ensuring the resilience of our networks and information systems requires collective and concerted national action that spans the whole of government, in collaboration with the private sector and individual citizens. For a decade, the United States has been fostering a culture of cybersecurity and an effective apparatus for risk mitigation and incident response. We continue to emphasize that systematically adopting sound information technology practices—across the public and private sectors—will reduce our Nation's vulnerabilities and strengthen networks and systems. We are also making steady progress towards shared situational awareness of network vulnerabilities and risks among public and private sector networks. We have built new initiatives through our national computer security incident response team to share information among government, key industries, our critical infrastructure sectors, and other stakeholders. And we continually seek new ways to strengthen our partnership with the private sector to enhance the security of the systems on which we both rely.

Strength Abroad. This model of defense has been successfully shared internationally through education, training and ongoing operational and policy relationships. Today, through existing and developing collaborations in the technical and military defense arenas, nations share an unprecedented ability to recognize and respond to incidents—a crucial step in denying would-be attackers the ability to do lasting damage to our national and international networks. However, a globally distributed network requires globally distributed early warning capabilities. We must continue to produce new computer security incident response capabilities globally, and to facilitate their interconnection and enhanced computer network defense. The United States has a shared interest in assisting less developed nations to build capacity for defense, and in collaboration with our partners, will intensify our focus on this area. Building relationships with friends and allies will increase collective security across the international community.

Deterrence

The United States will ensure that the risks associated with attacking or exploiting our networks vastly outweigh the potential benefits. We fully recognize that cyberspace activities can have effects extending beyond networks; such events may require responses in self-defense. Likewise, interconnected networks link nations more closely, so an attack on one nation's networks may have impact far beyond its borders.

In the case of criminals and other non-state actors who would threaten our national and economic security, domestic deterrence requires all states have processes that permit them to investigate, apprehend, and prosecute those who intrude or disrupt networks at home or abroad. Internationally, law enforcement organizations must work in concert with one another whenever possible to freeze perishable data vital to ongoing investigations, to work with legislatures and justice ministries to harmonize their approaches, and to promote due process and the rule of law—all key tenets of the Budapest Convention on Cybercrime.

When warranted, the United States will respond to hostile acts in cyberspace as we would to any other threat to our country. All states possess an inherent right to self-defense, and we recognize that certain hostile acts conducted through cyberspace could compel actions under the commitments we have with our military treaty partners. We reserve the right to use all necessary means—diplomatic, informational, military, and economic—as appropriate and consistent with applicable international law, in order to defend our Nation, our allies, our partners, and our interests. In so doing, we will exhaust all options before military force whenever we can; will carefully weigh the costs and risks of action against the costs of inaction; and will act in a way that reflects our values and strengthens our legitimacy, seeking broad international support whenever possible.

Development: Building Prosperity and Security

The United States will continue to demonstrate our conviction that the benefits of a connected world are universal. The virtues of an open, interoperable, secure, and reliable cyberspace should be more available than they are today, and as the world's leading information economy, the United States is committed to ensuring others benefit from our technical resources and expertise.

Our Nation can and will play an active role in providing the knowledge and capacity to build and secure new and existing digital systems, and in so doing, build consensus among states to behave as responsible stakeholders. Building capacity to realize these goals is not a short-term expenditure, but a wise long-term investment and a commitment on the part of our government for continued engagement.

Development Objective

The United States will facilitate cybersecurity capacity-building abroad, bilaterally and through multilateral organizations, so that each country has the means to protect its digital infrastructure, strengthen global networks, and build closer partnerships in the consensus for open, interoperable, secure, and reliable networks.

Building Technical Capacity

Access to networked technology is increasingly a basic need for development. Governments and industry have made a number of meaningful steps to enhance connectivity to end-users across un-served or underserved regions. International information infrastructures continue to mature and expand, providing more nations with the opportunity to access the global flow of information. The growth of the networks worldwide, and expansion of access to them, enriches the world community, yet also presents new challenges and opportunities for collaboration on issues of traditional and cybersecurity. Much of this capacity will result from private-sector investment, and the United States will work with governments and industry to build a climate friendly to those efforts, and in which they can be leveraged to address countries' core development needs.

Governments are a major determinant of whether this new connectivity produces positive outcomes or wastes its potential. Those states that have benefitted most from our capacity-building efforts are those that embrace technology to build prosperity and enhance social cohesion, rather than restrict access for the purposes of political control. For that reason, technical projects that the United States supports will by design enhance security and commerce, safeguard the free flow of information, and promote the global interoperability of networks.

Building Cybersecurity Capacity

Prosperity cannot be built on a foundation of fear and unreliability, and the United States is committed to helping build cybersecurity capacity alongside states' own technological development. Enhancing national-level cybersecurity among developing nations is of immediate and long-term benefit, as more states are equipped to confront threats emanating from within their borders and in turn, build confidence in globally interconnected networks and cooperate across borders to combat criminal misuse of information technologies. It is also essential to cultivating dynamic, international research communities able to take on next-generation challenges to cybersecurity.

Acknowledging that cybersecurity is a global issue that must be addressed with national efforts on the part of all countries, we will expand and regularize initiatives focused on cybersecurity capacity building—with enhanced focus on awareness-raising, legal and technical training, and support for policy development. Such programs must address more than purely technology issues; we will work with states to recognize the breadth of the cybersecurity challenge, assist them in developing their own strategies, and build capacity across the whole range of sectors—from network security and the establishment of Computer Emergency Readiness Teams (CERTs), to international law enforcement and defense collaboration, to productive relationships with the domestic and international private sector and civil society.

Building Policy Relationships

The United States' capacity-building assistance is envisioned as an investment, a commitment, and an important opportunity for dialogue and partnership. As countries develop a stake in cyberspace issues, we intend our dialogues to mature from capacity-building to active economic, technical, law enforcement, defense and diplomatic collaboration on issues of mutual concern. We will also facilitate relationships among countries developing cybersecurity capacity—using both regional fora and technical bodies possessing specialized expertise—and will continue to promote the sharing of best practices, lessons learned, and international technical exchanges.

III. Policy Priorities

The United States will continue to take action to help build and sustain open, interoperable, secure, and reliable networks at home and abroad, both for our citizens and others in the global community. Our approach is guided by the fundamental principles, driven by the overarching goal, and sustained by the policies outlined in this document—together they form the basis of the United States' international cyberspace strategy.

To fully realize this future in which cyberspace lives up to its potential for all, the United States Government organizes its activities across seven interdependent areas of activity, each demanding collaboration within our government, with international partners, and with the private sector. Taken as a whole, they form the action lines of our strategic framework.

For the many departments and agencies of the United States Government already engaged in these activities, they provide reinforcement to the important work already underway. For those developing implementation plans to carry out their specific responsibilities in cyberspace, they provide context and ensure unity of effort. The policy priorities outlined here call for and guide those specific actions, highlighting areas of past, present and future emphasis that demand concerted attention and resources at the national level.

Economy: Promoting International Standards and Innovative, Open Markets

To ensure that cyberspace continues to serve the needs of our economies and innovators, we will:

- **Sustain a free-trade environment that encourages technological innovation on accessible, globally linked networks.** Just as the free flow of information is critical to the functioning of our networks, free trade helps support innovation and market growth in the information age. The global embrace of the Internet can largely be traced to the spread of lower-cost and globally available computers and network technology. Competition in these markets drives innovation, while a free-trade environment enables manufacturers to keep prices competitive and standards high. Respecting the international standards of technology development and trade is an essential part of sustaining open markets, and enables leading-edge technology companies to rapidly deliver the benefits of their innovative products and services. Over the next few decades, the globalization of technology manufacturing will only increase, with substantial benefits for our networks and consumers. The United States will work to sustain that free-trade environment, particularly in support of the high-tech sector, to ensure future innovation.

- **Protect intellectual property, including commercial trade secrets, from theft.** The same global networks that power innovation also open up new avenues for industrial espionage and the theft of intellectual property and commercial information. Cyberspace can be used to steal an unprecedented volume of information from businesses, universities, and government agencies; such stolen information and technology can equal billions of dollars of lost value. Individual incidents often go unreported or undetected. Results can range from unfair competition to the bankrupting of entire firms, and the national impact may be orders of magnitude larger. The persistent theft of intellectual property, whether by criminals, foreign firms, or state actors working on their behalf, can erode competitiveness in the global economy, and businesses' opportunities to innovate. The United States will take measures to identify and respond to such actions to help build an international environment that recognizes such acts as unlawful and impermissible, and hold such actors accountable.

- **Ensure the primacy of interoperable and secure technical standards, determined by technical experts.** Developing international, voluntary, consensus-based cybersecurity standards and deploying products, processes, and services based upon such standards are the basis of an interoperable, secure and resilient global infrastructure. The public and private sectors must work together to develop, maintain, and implement these standards and support the development of international standards and conformity assessment schemes that prevent barriers to international trade and commerce. International cybersecurity standardization, and its voluntary and consensus-based processes, serves collective interests. They foster innovation; facilitate interoperability, security, and resiliency; improve trust in online transactions; and spur competition in global markets. The United States will foster collaboration between the public and private sector to ensure the promulgation of international standards-based requirements for products and services.

Protecting Our Networks: Enhancing Security, Reliability, and Resiliency

Because strong cybersecurity is critical to national and economic security in the broadest sense, we will:

- **Promote cyberspace cooperation, particularly on norms of behavior for states and cybersecurity, bilaterally and in a range of multilateral organizations and multinational partnerships.** An increasing number of international organizations are taking up cybersecurity and other cyberspace issues, and the United States continues to promote this important work, building cyberspace into their range of work to meet the needs of their varied memberships. We have worked to include relevant cyberspace issues on the agenda at the Organization of American States (OAS), the Association of Southeast Asian Nations (ASEAN) Regional Forum (ARF), the Asia-Pacific Economic Cooperation Organization (APEC), the Organization for Cooperation and Security in Europe (OSCE), the African Union (AU), the Organization for Economic Cooperation and Development (OECD), the Group of Eight (G-8), the European Union (EU), the United Nations (U.N.), and the Council of Europe, and to ensure that work is supported by an effective institutional framework. The United States will continue, in these and other fora, to consolidate regional and international consensus on key cyberspace activities, including norms. We will also look to fora that enable multi-stakeholder collaboration and consensus building, to further elaborate the Internet policy principles outlined in this document. We welcome the expansion of this work to geographic regions currently underrepresented in the dialogue—most notably Africa and the Middle East—to further our interest in building worldwide capacity.

- **Reduce intrusions into and disruptions of U.S. networks.** Unauthorized network intrusions threaten the integrity of economies and undermine national security. Agencies across the United States Government are collaborating, together with the private sector, to protect innovation from industrial espionage, to protect Federal, state, and local government networks, to protect military operations from degraded operating environments, and to secure critical infrastructure against intrusions and attacks—particularly those on energy, transportation, or financial systems, and the defense industrial base. The United States will pursue a broad international consensus of states that recognize the importance of respect for property and network stability, and will back up that conviction with our own and our partners' willingness to defend our networks from acts that would compromise them.

- **Ensure robust incident management, resiliency, and recovery capabilities for information infrastructure.** In an interconnected global environment, weak security in one nation's systems compounds the risk to others. No one nation can have full insight into the world's networks; we have an obligation to share our insights about our own networks and collaborate with others when events might threaten us all. As we continue to build and enhance our own response capabilities, we will work with other countries to expand the international networks that support greater global situational awareness and incident response—including between government and industry. The United States Government actively participates in watch, warning, and incident response through exchanging information with trusted networks of international partners. We will expand these capabilities through international collaboration to enhance overall resilience. The United States will also work to engage international participation in cybersecurity exercises, to elevate and strengthen established operating procedures with our partners.

- **Improve the security of the high-tech supply chain, in consultation with industry.** The operation of critical networks and information infrastructures depends on the assured availability of trustworthy hardware and software. Vulnerabilities in the supply chain can enable attacks on the integrity, availability, or confidentiality of networks and the data they contain. Exploitation of these vulnerabilities impairs economic performance and national security. The United States will work with industry and international partners to develop best practices for protecting the integrity of information systems and critical infrastructure. In this way, we will greatly enhance the security of the globalized supply chains on which free and open trade depend.

Law Enforcement: Extending Collaboration and the Rule of Law

To enhance confidence in cyberspace and pursue those who would exploit online systems, we will:

- **Participate fully in international cybercrime policy development.** The United States is committed to participating actively in discussions about how international norms and measures on cybercrime are developed bilaterally and multilaterally, in fora with proven expertise and a history of promoting effective cybercrime policies. These conversations will incorporate existing efforts, like how to extend the reach of institutions like the Budapest Convention. The United States will build these efforts upon the successful partnerships between national law enforcement agencies and the productive policy dialogues that we currently enjoy, cultivating a sense of responsibility among states joining this effort.

- **Harmonize cybercrime laws internationally by expanding accession to the Budapest Convention.** The United States and our allies regularly depend upon cooperation and assistance from other countries when investigating and prosecuting cybercrime cases. This cooperation is most effective and meaningful when the countries have common cybercrime laws, which facilitates evidence-sharing, extradition, and other types of coordination. The Budapest Convention on Cybercrime provides countries with a model for drafting and updating their current laws, and it has proven to be an effective mechanism for enhancing international cooperation in cybercrime cases. The United States will continue to encourage other countries to become parties to the Convention and will help current non-parties use the Convention as a basis for their own laws, easing bilateral cooperation in the short term, and preparing them for the possibility of accession to the Convention in the long term.

- **Focus cybercrime laws on combating illegal activities, not restricting access to the Internet.** Criminal behavior in cyberspace should be met with effective law enforcement, not policies that restrict legitimate access to or content on the Internet. To advance this goal, the United States Government works on a bilateral and multilateral basis to ensure that countries recognize that online crimes should be approached by focusing on preventing crime and catching and punishing offenders, rather than by broadly limiting access to the Internet, as a broad limitation of access would affect innocent Internet users as well. As the United States and our partners engage in dialogue and help build capacity among law enforcement organizations worldwide, we will integrate this approach, uniting protection of privacy, fundamental freedoms, and innovation with collaboration to combat crimes in cyberspace.

- **Deny terrorists and other criminals the ability to exploit the Internet for operational planning, financing, or attacks.** The United States has a variety of international capacity-building and training programs on cybercrime, helping law enforcement and legislators develop effective legal frameworks and expertise to investigate and prosecute terrorist and other criminal misuse of the Internet. Preventing terrorists from enhancing capabilities through "hackers for hire" and organized crime tools is an important priority for the international community, and demands effective cybercrime laws. The United States is committed to tracking and disrupting terrorist and cybercrime finance networks through technical tools and international cooperation frameworks such as the Financial Action Task Force.

Military: Preparing for 21st Century Security Challenges

Since our commitment to defend our citizens, allies, and interests extends to wherever they might be threatened, we will:

- **Recognize and adapt to the military's increasing need for reliable and secure networks.** We recognize that our armed forces increasingly depend on the networks that support them, and we will work to ensure that our military remains fully equipped to operate even in an environment where others might seek to disrupt its systems, or other infrastructure vital to national defense. Like all nations, the United States has a compelling interest in defending its vital national assets, as well as our core principles and values, and we are committed to defending against those who would attempt to impede our ability to do so.

- **Build and enhance existing military alliances to confront potential threats in cyberspace.** Cybersecurity cannot be achieved by any one nation alone, and greater levels of international cooperation are needed to confront those actors who would seek to disrupt or exploit our networks. This effort begins by acknowledging that the interconnected nature of networked systems of our closest allies, such as those of NATO and its member states, creates opportunities and new risks. Moving forward, the United States will continue to work with the militaries and civilian counterparts of our allies and partners to expand situational awareness and shared warning systems, enhance our ability to work together in times of peace and crisis, and develop the means and method of collective self-defense in cyberspace. Such military alliances and partnerships will bolster our collective deterrence capabilities and strengthen our ability to defend the United States against state and non-state actors.

- **Expand cyberspace cooperation with allies and partners to increase collective security.** The challenges of cyberspace also create opportunities to work in new ways with allied and partner militaries. By developing a shared understanding of standard operating procedures, our armed forces can enhance security through coordination and greater information exchange; these engagements will diminish misperceptions about military activities and the potential for escalatory behavior. Dialogues and best practice exchanges to enhance partner capabilities, such as digital forensics, work force development, and network penetration and resiliency testing will be important to this effort. The United States will work in close partnership with like-minded states to leverage capabilities, reduce collective risk, and foster multi-stakeholder initiatives to deter malicious activities in cyberspace.

Internet Governance: Promoting Effective and Inclusive Structures

To promote Internet governance structures that effectively serve the needs of all Internet users, we will:

- **Prioritize openness and innovation on the Internet.** The ability to distribute information efficiently on the Internet is at the very core of modern consumer, business, political, scientific, and educational activity. Governments around the globe recognize the value of the Internet; however, many of them place arbitrary restrictions on the free flow of information or use it to suppress dissent or opposition activities. The method and enforcement of these restrictions vary widely across countries, as do their justification, but we should not allow the Internet's governance or technical architecture to be reengineered to accommodate decisions that violate fundamental freedoms, or unnecessarily stifle innovation. Effective, inclusive Internet governance can help ensure acts grossly outside international norms of acceptable network management are not compounded by a technical or governance structure that would enable them. Preserving, enhancing, and increasing access to an open, global Internet is a clear policy priority. The United States will continue to advance these goals through a variety of engagements, including outreach to appropriate multi-stakeholder institutions and organizations, and to relevant intergovernmental and nongovernmental organizations.

- **Preserve global network security and stability, including the domain name system (DNS).** Given the Internet's importance to the world's economy, it is essential that this network of networks and its underlying infrastructure, the DNS, remain stable and secure. To ensure this continued stability and security, it is imperative that we and the rest of the world continue to recognize the contributions of its full range of stakeholders, particularly those organizations and technical experts vital to the technical operation of the Internet. The United States recognizes that the effective coordination of these resources has facilitated the Internet's success, and will continue to support those effective, multi-stakeholder processes.

- **Promote and enhance multi-stakeholder venues for the discussion of Internet governance issues.** The very architecture of the Internet embodies a mode of social and technical organization which is decentralized, cooperative, and layered. Each of these characteristics is fundamental to the benefits the Internet has brought. That architecture fuels the freedom of innovation that enables economic growth. It fuels the freedom of expression and association that enables social and political growth and the functioning of democratic societies worldwide. The United States stands firm in our conviction that when the international community meets to discuss the range of Internet governance issues, these conversations must take place in a multi-stakeholder manner; we will continue to support successful venues like the Internet Governance Forum, which embodies the open and inclusive nature of the Internet itself by allowing nongovernment stakeholders to contribute to the discussion on equal footing with governments.

International Development: Building Capacity, Security, and Prosperity

To promote the benefits of networked technology globally, enhance the reliability of our shared networks, and build the community of responsible stakeholders in cyberspace, we will:

- **Provide the necessary knowledge, training, and other resources to countries seeking to build technical and cybersecurity capacity.** The benefits of an interconnected world should not be limited by national borders. For over a decade the United States has helped bridge that gap, supporting a variety of programs to help other nations gain the resources and skills to build core capacities in technology and cybersecurity. Our goal is to help other states learn from our experience, and in particular to build cybersecurity into their national technical development. Because the needs are many and diverse, our programs range from supporting national capabilities for incident management; to building public/private partnerships; to enhancing control systems security; to drafting effective laws to investigate and prosecute cybercrime; to developing and implementing programs to raise cybersecurity awareness and build a national culture of cybersecurity. Our work has taken place bilaterally, through foreign assistance, as well in partnership with innovative public-private initiatives like the United States Telecommunications Training Institute. In recent years, we have helped make this work a priority at multilateral fora such as the OAS, APEC, and the U.N. The United States will expand these collaborations, work in-country to support private-sector investment in capacity, draw attention to this critical need, and work to build new collaborations in the coming years.

- **Continually develop and regularly share international cybersecurity best practices.** Today, nations no longer need to develop cybersecurity capacity exclusively through a process of trial and error. We have worked with dozens of other states and with numerous multilateral organizations to develop and share best practices designed to help states make wiser investments and develop more effective policies. The United States will continue to identify, develop, and refine best practices and technical standards in collaboration and close partnership with industry, and will expand our efforts to promote awareness of and access to them. We will further promote collaborative science and technology research to enhance cybersecurity tools and capabilities.

- **Enhance states' ability to fight cybercrime—including training for law enforcement, forensic specialists, jurists, and legislators.** Because criminal cases involving computer networks often involve evidence and targets located overseas, governments regularly rely on one another to provide often extensive technical and investigative assistance in matters relating to serious crime and national security. Criminal threats can originate from any connected country, and many countries need substantial help in developing the investigative capacities required to collaborate in such investigations. By providing training on these issues, we develop critical contacts and help promote law enforcement technical understanding. This engagement will increase the prospects for effective law enforcement cooperation and reciprocal assistance. The United States will continue to pursue this objective by providing training in numerous regions, continuing our work in Africa, and with APEC, ASEAN, G-8, and the OAS.

- **Develop relationships with policymakers to enhance technical capacity building, providing regular and ongoing contact with experts and their United States Government counterparts.** Over the last few years, a growing international community of policymakers focusing on cyberspace issues has provided new avenues for dialogue, launched new development and security initiatives, and strengthened countless bilateral relations. As we invest in developing countries' long-term future through technical and cybersecurity capacity-building, the United States is committed to building those assistance relationships into closer partnerships on issues of mutual concern. We have taken a lead role in convening fora, such as the Meridian Conference, which fosters collaboration on critical information infrastructure protection issues. The United States welcomes more states entering into the dialogue as they become increasingly invested in the future of cyberspace, and will work to build enduring relationships among our experts and policymakers.

Internet Freedom: Supporting Fundamental Freedoms and Privacy

To help secure fundamental freedoms as well as privacy in cyberspace, we will:

- **Support civil society actors in achieving reliable, secure, and safe platforms for freedoms of expression and association.** We encourage people all over the world to use digital media to express opinions, share information, monitor elections, expose corruption, and organize social and political movements, and denounce those who harass, unfairly arrest, threaten, or commit violent acts against the people who use these technologies. Such cultures of fear discourage others in the community from using new technologies to report, organize, and exchange ideas.

The same protections must apply to Internet Service Providers and other providers of connectivity, who too often fall victim to legal regimes of intermediary liability that pass the role of censoring legitimate speech down to companies. The United States will be a tireless advocate of fundamental freedoms of speech and association through cyberspace; will work to empower civil society actors, human rights advocates, and journalists in their use of digital media; and will work to encourage governments to address real cyberspace threats, rather than impose upon companies responsibilities of inappropriately limiting either freedom of expression or the free flow of information.

- **Collaborate with civil society and nongovernment organizations to establish safeguards protecting their Internet activity from unlawful digital intrusions.** Promoting cybersecurity among civil society and nongovernmental organizations helps ensure that freedoms of speech and association are more widely enjoyed in the digital age. Cybersecurity is particularly important for activists, advocates, and journalists on the front lines who may express unpopular ideas and opinions, and who are frequently the victims of disruptions and intrusions into their email accounts, websites, mobile phones, and data systems. The United States supports efforts to empower these users to protect themselves, to help ensure their ability to exercise their free expression and association rights on the new technologies of the 21st century.

- **Encourage international cooperation for effective commercial data privacy protections.** Protecting individual privacy is essential to maintaining the trust that sustains economic and social uses of the Internet. The United States has a robust record of enforcement of its privacy laws, as well as encouraging multi-stakeholder policy development. We are continuing to strengthen the U.S. commercial data privacy framework to keep pace with the rapid changes presented by networked technologies. We recognize the role of applying general privacy principles in the commercial context while maintaining the flexibility necessary for innovation. The United States will work toward building mutual recognition of laws that achieve the same objectives and enforcement cooperation to protect privacy and promote innovation.

- **Ensure the end-to-end interoperability of an Internet accessible to all.** Users should have confidence that the information they transmit over the Internet will be received as it was intended, anywhere in the world. Equally important is the expectation that under normal circumstances, data will flow across borders without regard for its national origin or destination. Ensuring the integrity of information as it flows over the Internet gives users confidence in the network and keeps the Internet open as a reliable platform for innovation that drives growth in the global economy and encourages the exchange of ideas among people around the world. The United States will continue to make clear the benefits of an Internet that is global in nature, while opposing efforts to splinter this network into national intranets that deprive individuals of content from abroad.

IV. Moving Forward

The benefits of networked technology should not be reserved to a privileged few nations, or a privileged few within them. But connectivity is no end unto itself; it must be supported by a cyberspace that is open to innovation, interoperable the world over, secure enough to earn people's trust, and reliable enough to support their work.

Thirty years ago, few understood that something called the Internet would lead to a revolution in how we work and live. In that short time, millions now owe their livelihoods—and even their lives—to advances in networked technology. A billion more rely on it for everyday forms of social interaction. This technology propels society forward, accomplishing things previous generations scarcely thought possible. For our part, the United States will continue to spark the creativity and imagination of our people, and those around the world. We cannot know what the next great innovation will be, but are committed to realizing a world in which it can take shape and flourish.

This strategy is a roadmap allowing the United States Government's departments and agencies to better define and coordinate their role in our international cyberspace policy, to execute a specific way forward, and to plan for future implementation. It is a call to the private sector, civil society, and end-users to reinforce these efforts through partnership, awareness, and action. Most importantly, it is an invitation to other states and peoples to join us in realizing this vision of prosperity, security, and openness in our networked world. These ideals are central to preserving the cyberspace we know, and to creating, together, the future we seek.